NUMBER 655

THE ENGLISH EXPERIENCE

ITS RECORD IN EARLY PRINTED BOOKS
PUBLISHED IN FACSIMILE

(ENGLAND)

HIS MAIESTIES LETTERS PATTENT FOR AN OFFICE.. CALLED THE PUBLICKE REGISTER FOR GENERALL COMMERCE

(LONDON), 1611

WALTER J. JOHNSON, INC.
THEATRUM ORBIS TERRARUM, LTD.
AMSTERDAM 1974 NORWOOD, N.J.

The publishers acknowledge their gratitude to
the Syndics of Cambridge University Library
for their permission to reproduce the Library's
copy, Classmark: Bb*.11.11[10], and page A_1 r from
the Library's copy, Classmark: Syn.7.61.199[3]

S.T.C. No. 9226

Collation: A-E^4, F^2

Published in 1974 by

Theatrum Orbis Terrarum, Ltd.
O.Z. Voorburgwal 85, Amsterdam

&

Walter J. Johnson, Inc.
355 Chestnut Street
Norwood, New Jersey
07648

Printed in the Netherlands

ISBN 90 221 0655 1

Library of Congress Catalog Card Number:
74-80219

The Publique Re-giſter for generall Commerce.

A TRVE TRAN-SCRIPT AND PVB-LICATION OF HIS MAIESTIES LETTERS PATTENT.

FOR AN OFFICE TO BE EREC-TED, AND CALLED THE Publicke Regiſter for generall *Commerce*.

Whereunto is annexed an Ouerture and explanation of *the nature and purport of the ſaid Office, for their better* vnderſtanding and direction that ſhall haue occaſion to vſe it, By Sir *Arthur Gorges*, Knight.

Printed at Britaine Burſſe for *Iohn Budge*, and are there to bee ſold at his Shop. 1611.

I R
HONI SOIT QVI MAL Y PENSE

IAMES by the grace of God, King of England, Scotland, France, and Ireland, defendor of the Faith, &c. To all to whom these Presents shall come, Greeting. Whereas all Trade and Commerce whatsoeuer amongst our welbeloued subiects, doth chiefly consist eyther in Buying and Selling, or Borrowing and Lending: And for that a great defect is daily found in the *policie of our State for want of ſome good, truſty and ready meanes of intelligence and intercourſe betweene our ſaid Subiects in that behalfe. By meanes wherof, many men oftentimes vpon occaſion of neceſsity and ſudden accidents, are inforced to put away and ſell landes, leaſes, or other goods and chattels, to great loſſe and diſaduantage for want of good and ready meanes to giue generall notice and publique intelligence of ſuch their intentions, to many that would (if they knew thereof) as willingly buy as the others would gladly ſell. For redreſſe and remedy whereof, our truſty and welbeloued ſeruants* Sir Arthur Gorges *and* Sir Walter Cope *Knights, Gentlemen of our priuie*

Chamber

Chamber, haue out of their carefull endeauours found out and deuised a most safe, easie and speedy way to the great aduancement, and helpe vnto generall Commerce, and Trade amongst our welbeloued Subiects; whereby to serue the turnes and occasions aswell of Borrowers as Lenders, and of Buyers as of Sellers, *by plaine and direct course of reciprocall intelligence, and interchangeable correspondency, for the ready notice and vnderstanding of one anothers minds.* We *in our Princely care and prouidence for the generall good of our people (which we much desire and tender) haue thought it fit, and iust that the said* Sir Arthur Gorges *and* Sir Walter Cope *should re-*

ceiue

ceiue from vs some fauourable approbation and lawfull authoritie, to establish, further and aduance this their industrious care and faithfull endeauours for the publique good; whereby much ease, comfort and benefit would vndoubtedly ensue to our Subiects of this our Realme of England and the Dominions of the same, in the knowledge and vse thereof. Know *ye therfore, that we reposing great trust and confidence in the prouidence, integrity and fidelity of the said* Sir Arthur Gorges *and* Sir VValter Cope, *for their honest, iust, and respectiue dealing herein: of our especial grace, certaine knowledge, and meere motion, haue giuen and granted, and by these presents for Vs, our Heyres and Successors,*

ceſſors, doe giue and graunt abſolute, ful and free licenſe, power and authority vnto the ſaid Sir Arthur Gorges and Sir VValter Cope, their Executors, Adminiſtrators and Aſſignes, and to their Deputy and Deputies, during the terme of one and twenty yeares, next and immediatly enſuing the date of theſe Preſents, to erect, ſet vp, keepe and maintaine in any places, Cities or Townes within our Realme of England and the Dominions of the ſame, where the ſaid Sir Arthur Gorges and Sir VValter Cope, their Executors, adminiſtrators or aßigns ſhal thinke moſt fit and conuenient, a publique Office, roome or place of reſort or repaire of people for the notice of Borrowing and Lending *of moneyes, and for the better knowledge of buying, ſelling or exchanging of lands, tenements or hereditaments, leaſes or any other goods or chattels whatſoeuer, which they or their Deputies ſhal thinke fit, and worthy to be entred and regiſtred; and to keepe one or more Kalender or Kalenders, Regiſter or Regiſters, for the regiſtring of all and ſingular ſuch lands, tenements, hereditaments, leaſes, wares, commodities, moneyes, or any other things or chattels, that ſhall by the meere motion or good liking of the owners themſelues, or their Factors for them be brought to ſuch Office and Offices, there to be entred and regiſtred, to be bought, ſold, morgaged, pawned, borrowed or lent:.*

lent: which Office in euery Citie, Towne or place where it ſhall be kept, ſhall be called by the name of The Publique Regiſter for generall Commerce. *And alſo that it ſhall and may be lawful, to and for the ſaid* Sir Arthur Gorges *and* Sir VValter Cope, *their Executors, adminiſtrators or aſsignes, and their deputie and deputies, of and in the ſaid Office from time to time, during the ſaid terme of one and twenty yeares, to keepe entries, to make ſearches, and to continue Regiſters in places conuenient, of all ſuch landes, tenements, hereditaments, leaſes, goods, chattels, profits, commodities or merchandizes whatſoeuer, which they the ſaid* Sir Arthur Gorges *and* Sir VValter

Cope, *their Executors, administrators or assignes, or their deputy or deputies, or any of them shall thinke worthy to be entred or registred in the said Office, and that in all things according to the true intent and meaning of these presents, without let, hinderance or interruption of any our louing Subiects whatsoeuer.* Yeelding *and paying therfore during the terme aforesaid, to Vs, our Heires and Successors, from the feast of Saint* Michaell *the Archangell next ensuing the date hereof, the yearely rent of forty pounds of lawfull money of England, payable at the two vsuall feasts, of our Lady, and Saint* Michaell, *by equall portions.* And *we do therefore by these presents for Vs, our Heyres*

and

and Successors, prohibite, forbid, and straightly charge and command all, and every our Subiects whatsoever, that none of them during the said terme of one & twenty years, by these presents granted, doe attempt, undertake or presume to imitate, erect or exercise any like publique Office or Register of intelligence, or entercourse for Trade and Commerce, upon paine of our indignation and high displeasure. Alway *provided, that no man bee compelled to make Entrie or Search in the said Office at all, nor when such Entry or Search shall be voluntarily made, shall pay any more for such Search or Entrie then shall please himselfe. And also that every man bee left to his owne election and free choyce, to vse any Scriuener, Broker, friend, seruant or Factor, or any other for any the causes aforesaid, in as free, liberall, and ample manner as they haue heretofore vsed and accustomed. And that euery man who for seruice or other causes knowne to himselfe, shall be vnwilling to vse his owne name in the proceedings hereof, may vse the name of his seruant, friend, Factor, or any other at his own will and pleasure.* Prouided *also, and our will and pleasure is, that the said* Sir Arthur Gorges *and* Sir VValter Cope, *or either of them, their, or either of their Executors, administrators, or assignes, shall at or before the feast of* Saint Michaell *the Archangell,*

gell,

gell, which ſhall be in the yeare of our Lord God, One thouſand, ſixe hundred and twelue, finde by experience that they ſhall not be able by the gains ariſing by reaſon of this Office, to defray the maine charges, expences and rent incident to the ſaid Office and Officers; in reſpect thereof ſhall bee minded vtterly to relinquiſh and giue vp the ſaid Office, and the exerciſe thereof, and all profite and benefite which thereof may ariſe or grow; and the ſame their or any of their intentions and minds therein, ſhall ſignifie vnto the Lord Treaſurer of England for the time being, vnder their handes and ſeales: That then theſe our Letters Patents, as to them or eyther of them, their or eyther of their Executors,

tors,

tors, administrators or assignes, or any of them which shall so signifie vnto the said Lord Treasurer of England for the time being, his or their mind and intention, vtterly to relinquish and giue vp the said Office and exercise thereof: and all profits and benefits, which thereof can or may arise or grow, shall be vtterly voide, and the said party or parties, so signifying as is aforesaid, their Executors and administrators to be from the time of such signification, discharged of the said yearely rent, before in these presents reserued, which should afterwards incurre and grow due; and the other party and parties clayming interest in the said Office, and other the premises by, and vnder the said Letters

ters

ters Patents, and not so signifying his or their minde and intention, vnder his or their hand and seale vnto the said Lord Treasurer of England for the time being, as is aforesaid; and their Executors, administrators and assignes, to stand from thenceforth sole Officer and Officers, and bee solely and onely chargeable towards Us, our Heyres and Successors, for and with the said yearely rent, in and by these Presents reserued; and for and with all other Conenants, and things, which according to the tenour and true meaning of these presents, ought to bee performed by the said Sir Arthur Gorges *and* Sir Walter Cope, *their Executors, Administrators or assignes, as is aforesaid.* · In witnesse *where-of we haue caused these our Letters to be made Patent. Witnesse our selfe at Westminster the fift day of March; In the eighth yeare of our reigne of England, France, and Ireland: and of Scotland the foure and fortieth.*

CARTVVRIGHT.

Per me de priuato Sigillo.

Examinatur per me Thomam Marten.

An

TO THE VNPAR-
tiall Reader.

IN matters of this nature, where a loy-al deſire hath endeauoured to performe an acceptable ſeruice to the Common-wealth; albeit the ſucceſſe ſort not in euery point, anſwerable to hope and expectation: yet this may ſtand for a maine comfort, that the intention is free from all publique offence, or priuate wronge, and therefore respects not the taxe or scandall of those malignant humours, which are accustomed to cast corrupt aspertions on honest designes. But because this proiect hath meerly and solely proceeded from mine own poore conceit and labour, (ayming specially at the aduancement of mutuall Commerce, the bond and sinewes of humane society) I haue thought good for the auoyding of all doubtfull construction or mistaking (which might blemish it in the birth) to deliuer it into the world, with a perspicuous Ouerture and direction concerning the

purport and vſes therof; becauſe they are but briefly touched, and in generall terms ſet downe in the Patent.

Arthur Gorges.

AN OVERTVRE AND EXPLANATION of the purport and vſe of this OFFICE.

F*Irſt, it is to be vnderſtood that this Office to be knowne and called by the name of* The publique Regiſter for generall Commerce,*is indeed but the very reſemblance of a*

publique

publique Market, whereunto all men may freely repaire, and resort to Trade and Traffique, without constraint or restraint, at their owne wils and pleasures. And as in Markets many Commodities are solde and met withall more easily, and better cheape in one selfe place, then by seeking vp and downe the Countrey for them with more trouble, and at dearer rates in seuerall places: so is it by the vse and benefite of this Office, which as a publique Market, is solely in his Maiesties power to authorise and priuiledge.

Secondly this publique Market is a place which all men may easily finde out and freely repaire vnto, for speedy meanes of helpe vpon any extremity

of

of forfeitures of bondes or counter-bonds, or danger of morgages of landes or goods by breach of dayes, when they know not suddenly where to prouide themselues of moneyes vpon good security; for that euery man (especially those that dwell in remote Shieres) is not acquainted with a Broker, nor knowes assuredly where to finde one to serue his turne vpon such vrgent occasion, or if he doe, it will be at a deare rate: which defect this Register shall be able to supply with very conscionable and moderate consideration; taking nothing of any, but what the parties themselues shall in reason thinke fit and competent for the seruing of their turnes, by the trauell of the Officers.

D *Thirdly,*

Thirdly, by instance to make it more plaine, such men as haue monies lying by them, that they would expose after eight, nine, or ten in the hundred may, if they please, make it knowne & haue it registred in this Office, wherby such as repaire vnto this Register to borrow, may also receiue notice & direction accordingly for such summes as they desire by assignment from the Officers, if the security be good and liked of.

Fourthly, by way of example also to expresse the vse and benefite of this Office for Buying and Selling: *Suppose that one hath lands or a house in Wilt-shiere that he would gladly sell, suddenly to raise monies, without too*

great

great losse, but knows not how to make it generally knowne to the world; and that there are some others as desirous to buy lands or a house in that shire at a reasonable rate, and cannot learne nor get knowledge of any such; yet by recourse to this Register they may easily informe themselues, and haue mutuall intelligence of one anothers mindes perfectly, and readily, albeit they dwell farre asunder; wheras now many times both parties are in care and cannot serue their turnes accordingly, vndergoing thereby much inconuenience, which this Office with great ease doth remedie.

Fiftly, the parties themselues that do either lend or borrow, buy or sell, may vse it with as great security & secrecy as they shall thinke fit by vsing other mens names then their owne, other places then their owne houses, and other friends or Factors for themselues if they be so disposed. So as by this free, plaine and easie course all men may rest satisfied, that there is no cunning intent by meanes of this Office, to defraude men, nor by the practice thereof to discouer mens wealths or secret estates, further then to themselues shall seeme good.

Sixthly, it is neyther desired nor meant that any should leaue their moneyes or goods that they would lend or sell, in the charge or custody of this Office, but in such places as the owners

ners themselues shall thinke most safe and conuenient; wherby such as would borrow or buy, may (if they please) by this Register be directed to those places or persons that would Lend or Sell to serue their turns mutually. And by this course all men may rest secure of the integritie and faire dealings that shall bee vsed concerning other mens monies and goods; which the Patentees seek not to haue the custody or exposing of for their priuate profit, but will rather auoide the accidents and dangers of such hazards, and losses, as might happen in so great a charge, by mis-reckoning, robbing, or many other waies, which we see so often happens vnto men in their owne priuate houses. And such a misfortune

tune might be both a disparagement to our credits by wrong constructions, and also a taxe and preiudice to the proceeding and reputation of this Office (which consisting meerely of the good opinion of the world for the iust dealing to be vsed therein) if it be found contrarie or corrupt, it cannot subsist, but will die in it selfe as vnprofitable.

Moreouer wheras many men dwelling far from London, are now inforced to repaire thither, to their great trauaile & charge to serue their turns many waies in these forenamed kinds; by the benefite and vse of this Office they shall be much eased. For it is intended (by the fauour of God) that in

seue-

seuerall places of this Realme this publique Register shalbe kept in some such chiefe Cities or Townes as shall bee found most conuenient to serue the Countries next about them in these affaires, and to holde correspondency with the Citie of **London.**

Againe, it is very certaine that diuers landes, leases, houses, woods and chattels, which haue beene solde priuately at very low rates, to the no little losse of the owners, would haue bin purchased at more indifferent prices, if they had bin more publiquely knowen: for it often falleth out, that those who would gladly buy such, do seldome or neuer get knowledge of the Sellers intent, till the things desired be past

ree-

recouery, to the preiudice of both parties: which defect this publique Register will be a great meanes to supplie.

Furthermore, there are a world of extraordinarie things, which are but few mens monies, and the owners themselues would gladly vent if they knew how, but lie concealed or in effect lost in the sale, because that diuers others, which would as willingly buy such at valuable prices, know not how or where to recouer them. But herein it is to be vnderstood that no commodities of base qualitie or inferiour values shall be entred into this Register, but such as the Officers shall thinke fit to be dealt in, without disparagement of the place.

Be-

Besides, it often falleth out that men vpon extremity are driuen to pawne things of good value for smal summes, to serue their turnes with monies, vntill they can sell them outright; wherin they vse the means of Scriueners or Brokers, in whose hands they leaue the said pawnes, setting very reasonable prices on them; allowing also a reward: and yet notwithstanding it is often seene and spoken of, that many Scriueners or Brokers, to whom such things haue beene committed in trust, would neuer vent them at the owners prices, except they might make a greater surplusage to themselues then the rate limited. And by that meanes many haue beene kept in want of such reliefe, as their owne goods would af-

foord

foord them; whereby at last their pawnes haue eaten themselues out in vsury, or at least haue beene solde to great losse, the owners wanting means to redeeme them: to which grieuance this publique Register may yeelde a comfortable remedy.

Moreouer, whereas many conscionable men would willingly expose their monies for eight or nine in the hundred, but cannot make it knowne, and therefore doe put their stockes into the hands of Brokers or Scriueners to imploy for them: so it is, that those Brokers doe for the most part take ten in the hundred for those monies which the owners doe affoord at a cheaper hand; thereby wronging both

the

lender and borrower: which inconue-nience this publique Register would easily remedy, by giuing all men safe and ready meanes to expose their own monies; besides the assurance neuer to be deceiued in their trust, which now many times happens when Brokers or Scrieueners do chance to fall banck-rout.

Furthermore, whereas there is daily occasion of returning and pay-ing of monies from one part of the Kingdome to another, by meanes of Trade and Commerce: and therein is found a great defect for want of some safe and speedy meanes of dire-ction, for interchangeable corrcspon-dency in that behalfe (because that

 men

men are now inforced eyther to conuey those monies that they are to pay or receiue, by Carriers, Clothiers, or seruants, who are many times robbed; or else men are driuen to be at extraordinary cost, for the returnes of monies, by letters of exchange) by the vse of this publique Register men may (if they please) greatly ease and remedy both these inconueniences. And to expresse the manner of doing it by example (which is the plainest way of instruction,) Suppose that if any Gentleman, Merchant or Tradesman of the North, vpon occasion of busines, being come to London, giues notice there vnto this Register, that he can make present payment of monies in Yorke at a certaine day, to serue the

turnes

turnes of any Londoners, or others in that place, if they will deliuer the like summes in London; and so accordingly may any Londoners or others search the Register to that or the like intent, whereby to giue himselfe knowledge of any interchangeable correspondency, for payments or receits at Yorke. And in this sort may men mutually vnderstand one anothers mindes for Trade and Commerce, and with no lesse facility and security serue their turnes one of another in matters of this kinde, throughout all those Cities of this Kingdome, where this publique Register shall be erected and kept, which will be free for all men to repaire vnto, at their pleasures. But herein it is to be vnder-

derstood, that this publique Register shall not be made so vulgar, as to bee kept in euery chiefe Citie or Shiere-towne within the Realme, but in those necessary and commodious places that shall be found most fit for the publicke good, and to performe those beneficiall Offices that are promised and may be expected by the vse of this publique Register for the aduancement of generall Commerce.

Againe, to giue those men that would borrow full satisfaction, how they shall be assured when they come to this Register for direction in that behalfe, and haue giuen the Officer such gratuity, as in reason and out of his owne free will he hath thought

fit,

ſit, in regard of the note or direction deliuered to ſerue his turne, for the ſummes that vpon good ſecurity he would take vp, and not be diſappointed nor loſe his reward giuen: If it ſo fall out (for ſuch accidents may happen) that either the party that ſhould lend this money be from home, or elſe how otherwiſe ſo diſpoſed of his ſtocke, (ſince the intelligence giuen to this Regiſter) that he cannot inſtantly ſerue the turne of him that would borrow; let that party againe returne to the Office, and he ſhal either preſently haue another note, whereby to ſerue his turne, or elſe he ſhall receiue the reward that he hath giuen the Officer in that behalfe: Whereby may plainly appeare, that there ſhall be no

ſhifts

shifts vsed eyther to frustrate men of their desired purpose and hope in this Office, or trickes to draw rewards for that, which by the vse of the Office, or trauell of the Officer is not merited. And in this point I haue thought fit to giue good caution; both for that such chances may sometimes fall out: and besides it is not vnlikely, but that this obiection may be made. And therefore very pertinent to the purpose, and to auoyde doubts, plainely to set downe an answer, and certaine resolution, what men may depend vpon, and how they shall be dealt withall in such cases, thereby to auoyde all scruple. And as in matters of borrowing, so vpon occasions of buying, the like course and restitution shall be vsed.

But

But because amongst all these Ouertures and relations, there hath not beene any mention made, either when or where this publique Register shall be erected and kept; the which it may be presumed men will expect should haue beene in this discourse related, touching that point, it is very requisite this much to say: that the Office is of such a nature, as that it will require many things necessary for the complete ordering and fitting thereof, as also no lesse care and circumspection in the choice of such Officers and Clerkes, as shall be answerable and sutable to that integritie and iust dealing that is promised by the Patent, in the behalfe of the Patentees. Besides, to finde out and fashion a conuenient place fit for this Office could not be done suddenly, nor in hast, considering that our Patent was not past the Seale little more then a moneth since; and before that time the Patentees not assured whether it should passe, or not: and therefore no reason for them to make preparation for that which lay not in their owne powers to dispose of. Notwithstanding if it so fall out, that the proiect (in manner as it is expressed) proue acceptable to the world, in regard of the publique good, which is the true scope and end thereof, this may suffice, that all expedition shalbe vsed for the erecting thereof, and sufficient notice giuen by some publique manner of intelligence in what place it shal be kept.

Lastly,

Lastly, whereas his excellent Maiestie according to the custome of former times, and of all well policed States, hath in his Princely dispositi on pleased to giue grace and preferment to diuers honest Proiects and profitable inuentions, which haue truly tended to the publique good, without wronging the particular right of any; and in that regard hath also vouchsafed to grant his Letters Patent for the sole priuiledge and practise of this new inuented Register to the Patentees, if any vndutifull or malignant spirit for a selfe priuate gaine, will expostulate the iustnesse of this proceeding, it is thus truely and briefly answered, That it is in all equity as honourable for his Maiesty,

to

to prouide as good meanes for the ease and benefite of his people in generall (without any abridgement of former liberty) as it is lawfull or iust for any priuate Scriuener, Broker, or other, to prouide a course to aduance his owne particular estate and profite, which is often done to the detriment of many. And now that the nature, vse and scope of this publique Register is plainly and truly manifested, I leaue all men to their owne sense and liking, and the successe hereof to the eternall guider of all things.

FINIS.